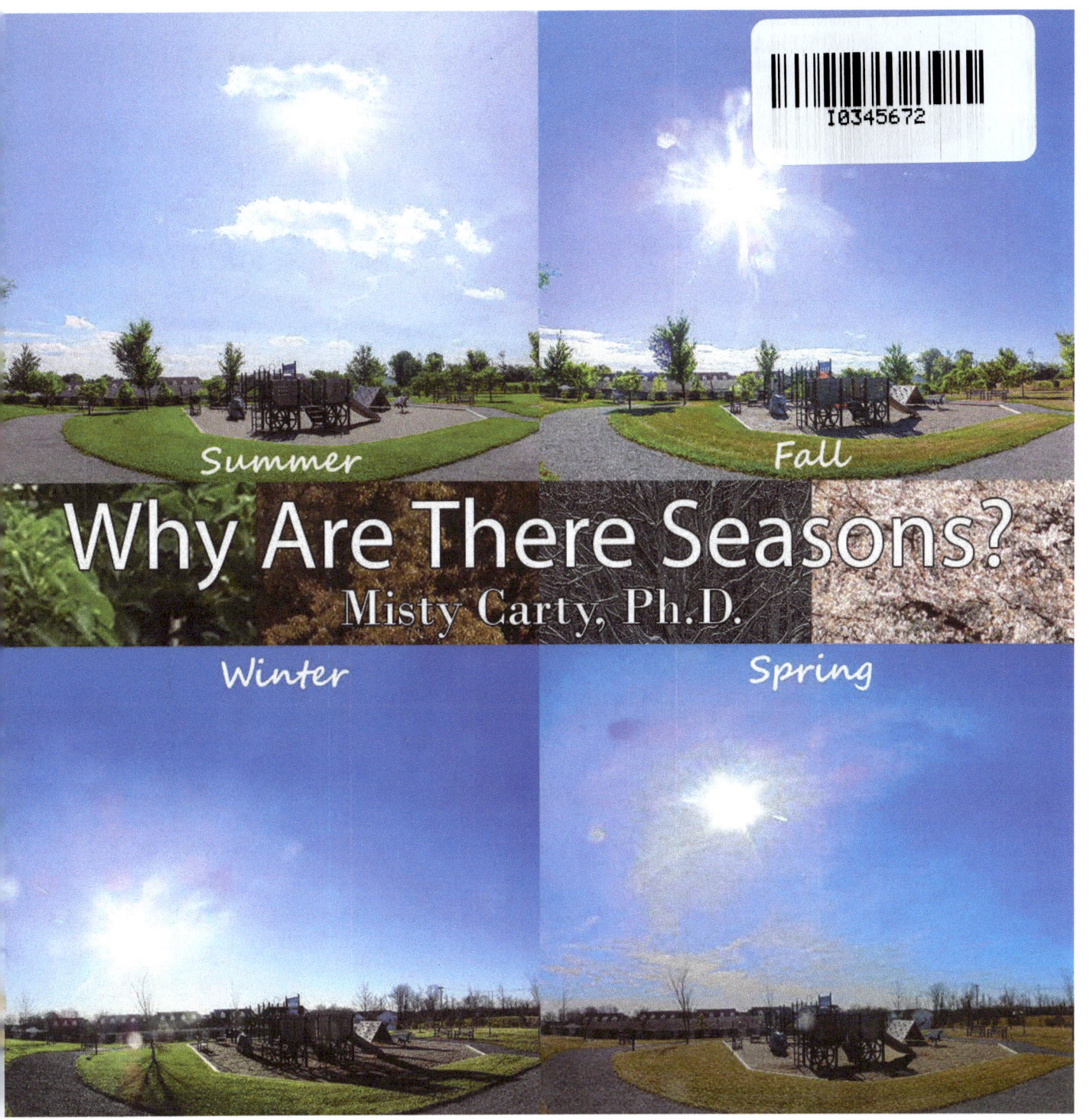

There are four seasons in a year: Summer, Fall, Winter, and Spring.

Summer is marked by hot days, green grass and green leaves on trees.

Fall brings cooler weather and colorful leaves.

Winter has cold days, bare trees, and snow.

Spring brings warmer weather and lots of beautiful flowers.

But, **WHY** do we have seasons?

Summer

Fall

Winter

Spring

We have seasons because the Earth leans to the side; it is *TILTED*!

The Earth spins around once in a day. As it spins, the Earth is tilted at an angle.

The Earth also moves around the Sun once in a year. As it moves around the Sun, the direction of the Earth's tilt changes – towards or away from the Sun.

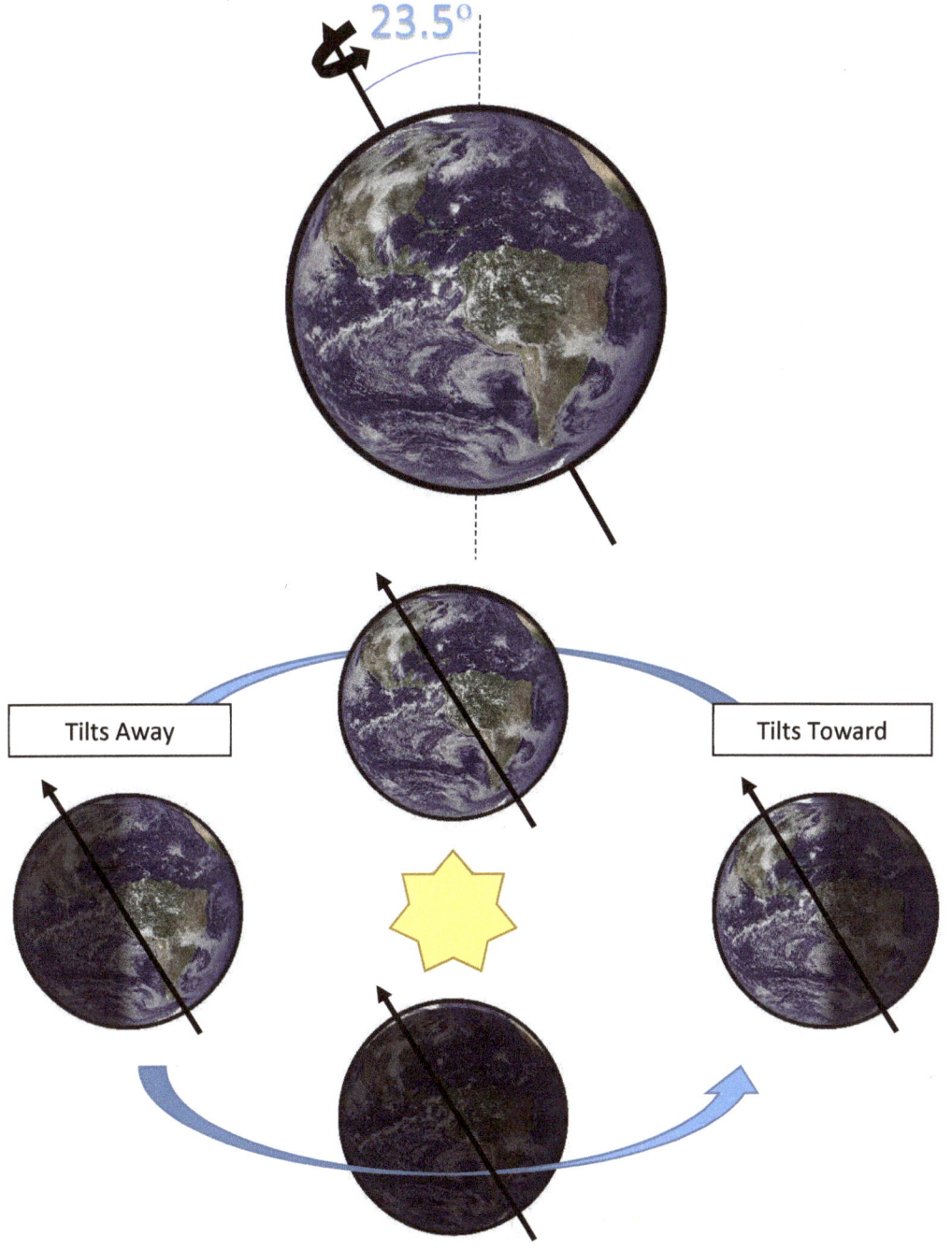

The Summer Solstice

On the first day of Summer, the Earth tilts the most towards the Sun. The Sun appears at its highest in the sky and casts the shortest shadows. The day is the longest. The night is the shortest.

When the Sun is at its highest, the Earth receives the most direct sunlight. This makes Summer the hottest season of the year.

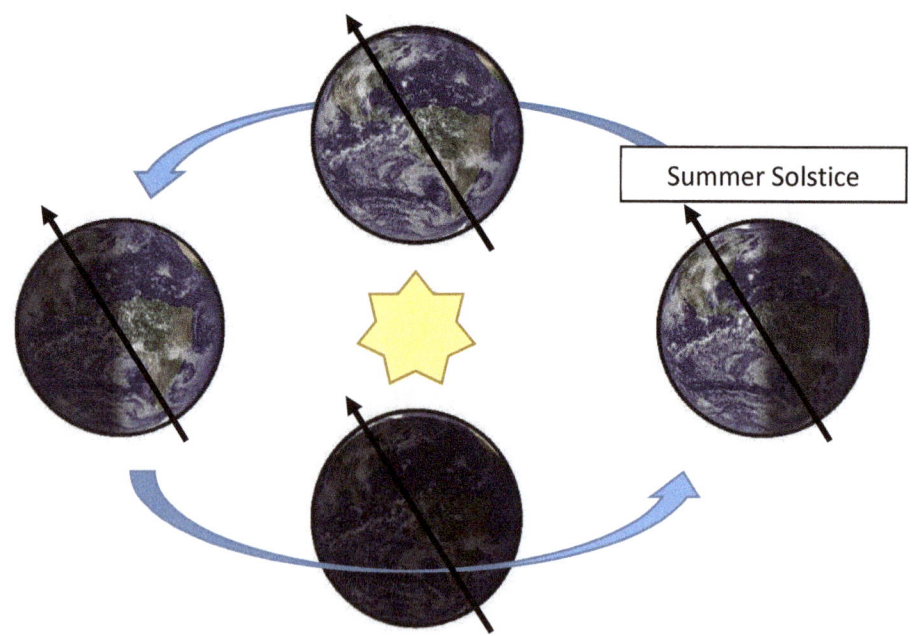

As Summer continues, the Sun begins to dip in the sky. The shadows are a little longer. The days are a little shorter. The nights are a little longer.

The Fall Equinox

On the first day of Fall, the Earth is neither tilted toward nor away from the Sun. This makes the day and the night the same length.

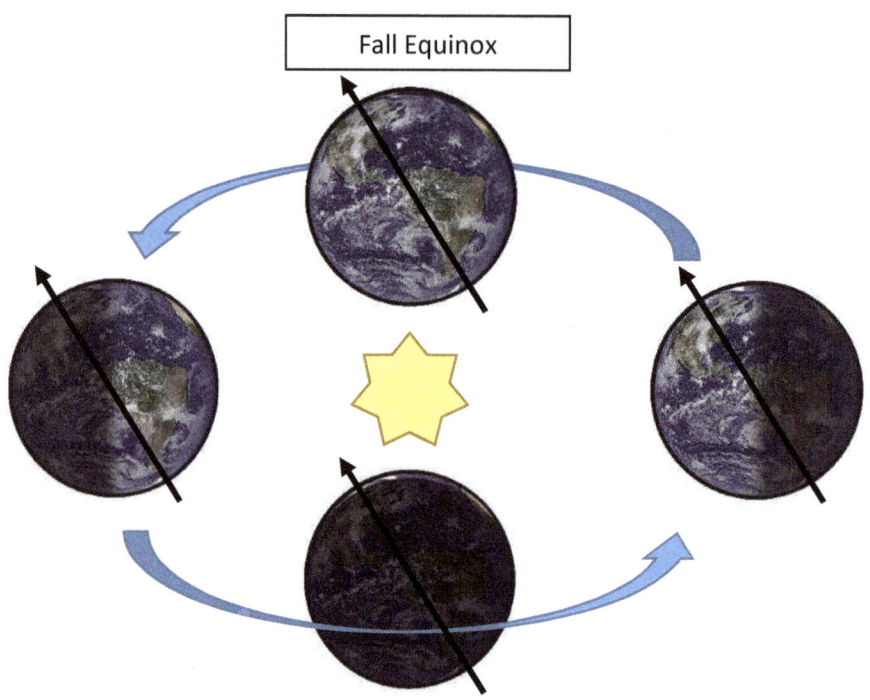

As Fall continues, the Sun dips even lower in the sky. The shadows are even longer. The days are now shorter than the nights.

The Earth receives a little less direct sunlight each day and begins to cool. This makes the leaves on the trees change color and fall.

The Winter Solstice

On the first day of Winter, the Earth tilts the furthest from the Sun. The Sun appears at its lowest in the sky and casts the longest shadows. The day is the shortest. The night is the longest.

When the Sun is at its lowest, the Earth receives the least amount of direct sunlight. This makes Winter the coldest season of the year.

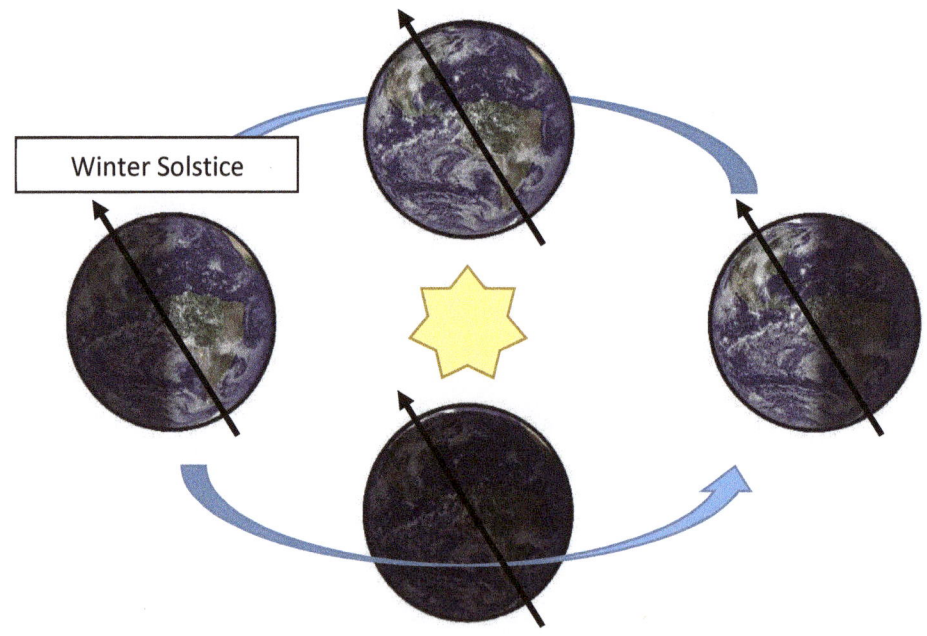

As Winter continues, the Sun begins to rise higher in the sky. The shadows begin to shorten. The days grow a little longer. The nights a little shorter.

The Spring Equinox

On the first day of Spring, the Earth is neither tilted toward nor away from the Sun. This makes the day and the night the same length.

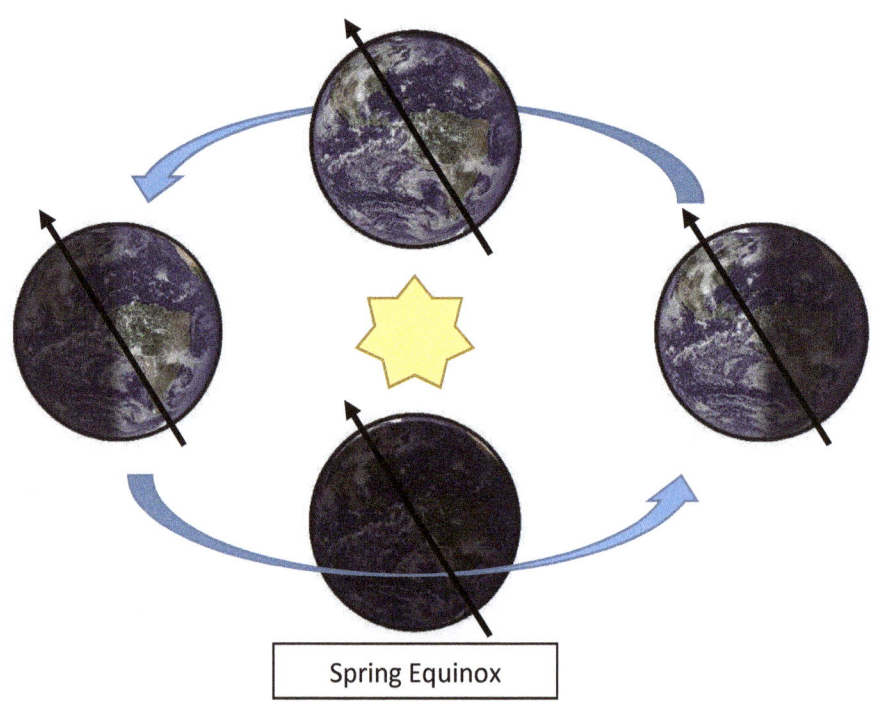

Spring Equinox

As Spring continues, the Sun rises even higher in the sky. The shadows are even shorter. The days are now longer than the nights.

The Earth receives a little more direct sunlight each day and begins to warm. This warming brings blossoms to the trees and flowers on the ground.

A year has passed and it is the Summer Solstice again. The Sun returns to its highest point, starting another yearlong trek across the sky.

Summer | Fall

During the year, the tilt of the Earth - towards or away from the Sun - changes the amount of direct sunlight it receives.

Winter | Spring

This is why we have seasons!

For Parents and Educators:

Introducing science to young children is fun! As any parent knows, kids are little scientists; experimenting and interacting with the world around them every day. Already curious, children love furthering their vocabulary and knowledge about the objects and actions they experience.

In this book, *Why Are There Seasons?*, your child will learn what causes the seasons. First, the seasons are presented in descriptive form; this connects the environmental changes your child experiences throughout the year to the corresponding season. Next, the characteristics of Earth's rotation on its axis and its revolution around the Sun are discussed; the physical aspects that lead to the Earth having seasons. Then, the seasons are illustrated two ways: the position of the Sun, as your child would see it, on the sky and Earth's position in its orbit around the Sun. Both perspectives are presented- our view from Earth and the view from space – to illustrate how the tilt of the Earth leads to varying amounts of direct sunlight throughout the year.....the reason we have seasons!

As you read this book with your child, please use the following information to deepen your science experience.

Our View of the Sun

To capture the Sun's change in position on the sky, the images were taken 1) in the same location and 2) at the same time of day, 4pm EDT and 3pm EST. They were photographed at Dowden's Ordinary Park, in Clarksburg, MD from June 2014 through September 2015.

As you read through the seasons, look for the following:

1) The change in height of the Sun. It is high in the sky at Summer, and moves lower until it reaches its lowest point in the sky at Winter. Then it treks back, moving higher and higher until Summer again.
2) The change in north-south position of the Sun. At Summer, the Sun is furthest north in the sky, setting in the northwest at night. At Fall and Spring, the Sun is due west. At Winter the Sun is furthest south, setting in the southwest at night.
3) The change in length of the shadows. In Summer, the shadows are very short and grow longer until Winter. Then they begin to shorten again from Winter until Summer.
4) The change in angle of the shadows. As the Sun moves from the northwest at Summer to the southwest at Winter, the angle of the shadows change as well.

All of these observations are due to the tilt of the Earth!

The Earth

The top image on page 5 illustrates the tilt of the Earth. The axis, about which the Earth spins, is tilted by about 23.5° from vertical.

The bottom image on page 5 illustrates the revolution of the Earth around the Sun. The Earth completes one revolution in a year. The tilt of the Earth points towards and away from the Sun as it revolves. The Northern Hemisphere is tilted the most toward the Sun on the Summer Solstice; it is tilted the most away on the Winter Solstice. On the Spring and Fall Equinoxes, the Earth is not tilted towards or away from the Sun - this makes the length of the day and the length of the night the same!

Image Credit: NASA GOES

Taking it Further

Observing changes due to the Earth's tilt is something everyone can do! You can start at any time during the year: beginning at either of the solstices or equinoxes can be a fun way to begin your experiment. Pick a place with a clear view of the horizon and a structure with a fixed height (or use your own, like a yard stick). Pick a time of day that you will observe; it is important that your observations occur at the same time of day, taking daylight savings time into account. Make your observations! Note the position of the Sun (be careful not to look directly at the Sun, the appearance of trees and plants, the characteristics of the weather, and the length of shadows. Collect your observations over a year and discover how the tilt of the Earth causes seasons!!!

Glossary

Equinox – the day the Earth is tilted neither towards nor away from the Sun. The length of the day and the night are the same. An equinox occurs twice in a year; in the Spring and the Fall.

Summer Solstice – the day the Northern hemisphere receives the most amount of direct sunlight. The Northern hemisphere is tilted the most toward the Sun. The sun appears at its highest in the sky.

Tilt – the amount an object leans to the side.

Winter Solstice – the day the Northern hemisphere receives the least amount of direct sunlight. The Northern hemisphere is tilted the most away from the Sun. The Sun appears at its lowest in the sky.